FLOAT AND SINK

by Robin Nelson

Lerner Publications Company · Minneapolis

What **floats?**

What **sinks?**

A duck floats.

A rock sinks.

A feather floats.

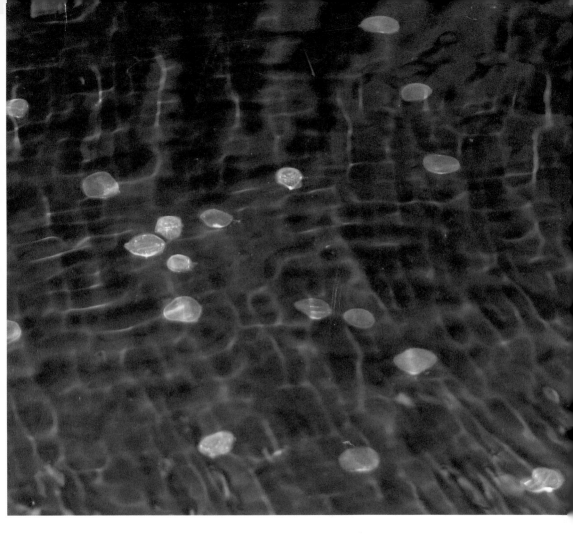

A penny sinks.

A boat floats.

An **anchor** sinks.

A ball floats.

A spoon sinks.

A **bobber** floats.

A **hook** sinks.

Bubbles float.

Soap sinks.

Can you float?

Can you sink?

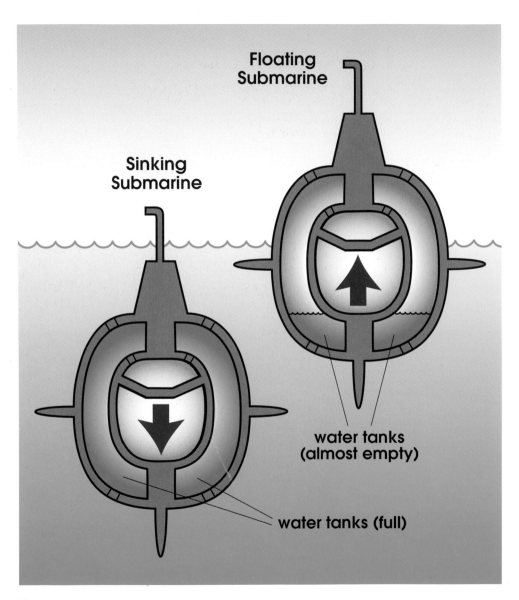

Floating Submarine

Sinking Submarine

water tanks (almost empty)

water tanks (full)

18

How does a submarine sink and float?

Sinking

Submarines have big tanks. The submarine lets water into the tanks. It lets in water until it is heavy enough to sink.

Floating

The submarine pumps the water out into the sea. It gets lighter and rises to the surface. The submarine floats.

Float and Sink Facts

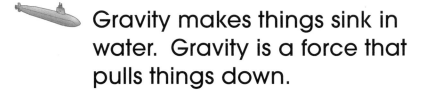 Gravity makes things sink in water. Gravity is a force that pulls things down.

An object floats if it is lighter than the water under it. The object floats because the water pushes up against it.

A small, heavy object will usually sink.

A large, light object will usually float.

A heavy ship can float because its weight is spread out over a large area. That way, more water is pushing up against the boat to keep it floating.

It is easier for an object to float in saltwater than in freshwater.

An iron ball would take more than an hour to sink to the bottom of the deepest part of the ocean.

Glossary

 anchor – something heavy that keeps a boat in one place

 bobber – a floating ball on a fishing line

 floats – sits on top of water

 hook – a curved piece of metal used to catch fish

 sinks – goes down under the water

Index

The photographs in this book are reproduced through the courtesy of: PhotoDisc Royalty Free by Getty Images, cover, pp. 2, 16, 22 (middle); © Todd Strand/Independent Picture Service, pp. 3, 4, 6, 7, 11, 14, 15, 22 (bottom); © Elwin Trump, p. 5; Stockbyte Royalty Free, p. 8; © Ken Hoppen, pp. 9, 22 (top); Corbis Royalty Free, pp. 10, 13, 17, 22 (second from bottom); © Russell Graves, pp. 12, 22 (second from top).

Illustration on page 18 by Laura Westlund.

Lerner Publications Company
A division of Lerner Publishing Group
241 First Avenue North
Minneapolis, MN 55401 USA

Website address: www.lernerbooks.com

Library of Congress Cataloging-in-Publication Data

Nelson, Robin, 1971–
    Float and sink / by Robin Nelson.
        p.   cm. — (First step nonfiction)
    Includes index.
    Summary: An introduction to the difference between floating and sinking.
    ISBN: 0–8225–5135–7 (lib. bdg. : alk. paper)
    1. Floating bodies—Juvenile literature. [1. Floating bodies.] I. Title. II. Series.
QC174.5.N45 2004
532'.25—dc22   2003013882

Manufactured in the United States of America
1 2 3 4 5 6 – DP – 09 08 07 06 05 04